# Home Maths Ages 11

Anita Straker

**CAMBRIDGE**
UNIVERSITY PRESS

Ask an adult to time you.

You need pencil and paper. Write only the answers.

| | | | |
|---|---|---|---|
| **1** | ⅓ of ½. | **8** | 927 ÷ 9. |
| **2** | Four cubed. | **9** | Write 580 centimetres in metres. |
| **3** | 25% of £185. | **10** | (48 + 32) ÷ 8. |
| **4** | 48 × 9. | **11** | Add 12, 13, 8, 9 and 5. |
| **5** | 5006 – 3990. | **12** | 60 × 29. |
| **6** | 1.6 x 1000. | **13** | ⅝ + ¼ . |
| **7** | 3.6 + 2.4. | **14** | 17.6 – 3.25. |

**2**

# Target fractions

Do this by yourself.

You need pencil and paper.

Each time, use three different fractions from this set.

Copy and complete these.

1. ... + ... + ... = $1\frac{11}{12}$

2. ... + ... + ... = $2\frac{2}{15}$

3. ... + ... + ... = $2\frac{1}{4}$

4. ... + ... + ... = $2\frac{1}{20}$

5. ... + ... + ... = $2\frac{3}{10}$

Add three fractions
Work systematically

# Shipwreck

Play this with a partner.

| S | H | I | P | W | R | E | C | K |
|----|----|----|----|----|----|----|----|----|
| 10 | 21 | 32 | 43 | 54 | 65 | 76 | 87 | 98 |

Each letter has a value.

Add up the numbers in the words below.

Your partner should check and say if you are right.

| | | | | | | | |
|---|---|---|---|---|---|---|---|
| **1** | S H E | **5** | H I S | **9** | H I S S |
| **2** | P I E | **6** | P E W | **10** | W H I P |
| **3** | H E R | **7** | P I P | **11** | C R E W |
| **4** | E W E | **8** | S E W | **12** | W I S H |

Think of more words using the letters of SHIPWRECK.

Ask your partner to work out what they are worth.

You must say if your partner is right.

Add three or four two-digit numbers
Use knowledge of addition facts

# Multiples

Play with a partner.

You need pencil and paper and three dice
(but one dice will do).

Take turns to roll all three dice (or roll one dice three times).

Use the numbers you roll to make a three-digit number.

With 3, 5 and 2, say, you could make any of these numbers:

## 325  352  253  235  532  523

If your number is a multiple of 3, score 3 points.

If it is a multiple of 8, score 8 points.

If it is a multiple of 11, score 11 points.

352, say,  would score 19 points as it is a multiple of 8 and of 11.

If you can't make any of these multiples, wait for your next turn.

The winner is the first player to get 100 points.

### Change the rules

a.　Make multiples of 4, 5 and 12.

b.　Make multiples of 3, 7 and 8.

c.　Make multiples of 2, 9 and 13.

Use knowledge of number facts and times-tabl
Apply tests of divisibility and practise division

Ask an adult to read you these.
You need pencil and paper.  Write only the answers.

**1**   Write 0.12 as a fraction.

**2**   Five sixths of 24.

**3**   43 times 4.

**4**   13 plus 19 minus 18.

**5**   18 multiplied by 25.

**6**   12 subtract 20.

**7**   Multiply 2.5 by 1000.

**8**   Write 580 centimetres in metres.

**9**   Which is more: 1 kilogram or 1 pound?

**10**  Ten thousand subtract five hundred.

**11**  Double the sum of 49 and 36.

**12**  Add together 7, 9, 5 and 3.

**13**  What is the cube root of 8?

**14**  Write 1.5 as a percentage.

You need pencil and paper.  Write only the answers.

**1**   What is the difference in temperature between 18°C and −9°C?

**2**   In 15 hours a car travels 645 miles.  What is its average speed?

**3**   Put these in order, smallest first:  ½,  1½,  2,  ¼,  1¾.

**4**   Divide 80 by 100.  Write the answer as a decimal.

**5**   85cm is cut off 2.5 metres of wood.  How much is left?

**6**   How many prime numbers lie between 50 and 60?

**7**   What must you multiply 13 by to make 169?

**8**   How many edges has a hexagonal prism?

**9**   Round 94.38 to one decimal place.

**10**  A meal for 4 people cost £13.80.
     What was the average cost?

# Fractions

Two or three people can play.

You need two dice.

Each player needs pencil and paper.

Take turns to roll the dice.

If the numbers are the same, roll again to get two different numbers.

Use your two numbers to make two fractions.

If you roll, say, 4 and 5 you would make ⁵⁄₄ (or 1¼) and ⁴⁄₅.

Write your two fractions and find their total.

The winner is the first to get 11 different totals.

> **Change the rules**
>
> Subtract the two fractions.
>
> Aim to make 11 different differences.

Add or subtract fractions

Ask an adult to time you.

You need pencil and paper.  Write only the answers.

**1**  3.9 + 8.34.

**2**  9 × 99.

**3**  42 – 19 + 24.

**4**  ½ × ⅕.

**5**  20.8 ÷ 4.

**6**  5.6 × 1000.

**7**  (8 – 3) × (13 – 4)

**8**  Find the average of 65 and 195.

**9**  Find the change from £20 for £15.61.

**10**  What fraction of 3cm is 5mm?

**11**  Subtract 10 from 708.

**12**  0.3 × 6.

**13**  £6.38 + £9.22.

**14**  Approximately, what is 13.1 × 1.9?

# Square puzzles

Do this by yourself.

You need pencil and paper and 24 matches or cocktail sticks.

24 matches can be arranged like this to make 9 equal squares.

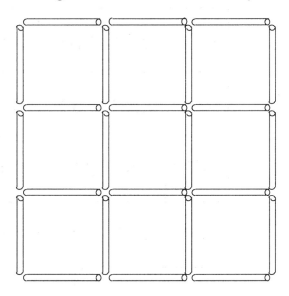

Find ways of arranging 24 matches:

**1**    as 4 squares, not necessarily equal;

**2**    as 5 squares, not necessarily equal;

**3**    as 6 equal squares;

**4**    as 7 equal squares;

**5**    as 7 squares, not necessarily equal;

**6**    as 8 equal squares.

Sketch what you did in each case.

Now ask your family to try.

ecognise different shapes made of squares
hink logically and visualise rearrangements

# Four in a line

Play with a partner.

Share a pencil and paper.

Draw a grid like this.

| 0.1 | 0.88 | 0.6 | 0.55 | 0.3 |
|------|------|------|------|------|
| 0.25 | 0.22 | 0.125 | 0.33 | 0.166 |
| 0.8 | 0.625 | 0.5 | 0.875 | 0.9 |
| 0.833 | 0.44 | 0.375 | 0.11 | 0.75 |
| 0.4 | 0.66 | 0.7 | 0.77 | 0.2 |

In turn, choose any two of these numbers.

Divide the smaller number by the larger.

If the answer is on the grid, mark it with your initial.

If the answer is not on the grid, or is already marked,
wait for your next turn.

The winner is the first to get four of their initials in a straight line.

The line can be in any direction.

Recognise the equivalence of fractions and decimals
Work out a strategy

You need pencil and paper. Write only the answers.

**1**   Which is closest to 4 litres:  4 pints, 7 pints, 14 pints or 17 pints?

**2**   How many different two-digit numbers can be made with 1, 2, 3, 4?

**3**   Sue doubles a number, adds 9, and gets 21.  What is the number?

**4**   How many rows of 25 chairs are needed to seat 320 people?

**5**   Parveen spent £13.26.  What was her change from £15?

**6**   What are the next three numbers:  2,  4,  8,  16,  … ?

**7**   It is 21:23.  What time will it in 4½ hours?

**8**   How many vertices has a tetrahedron?

**9**   Write $^{59}/_{8}$ as a mixed number.

**10**  Henry gallops 2½ miles in 5 minutes.
     What is his average speed in m.p.h?

Ask an adult to read these to you.
You need pencil and paper.  Write only the answers.

**1**   Divide 56 000 by 100.

**2**   0.7 times 9.

**3**   Add 10 to 993.

**4**   Six cubed.

**5**   Double 246.

**6**   48 plus 37.

**7**   90 divided by 5.

**8**   Multiply 7 by 300.

**9**   Subtract 67 from 132.

**10**  What is the cost of sixteen 7p stamps?

**11**  Write 6 out of 10 as a percentage.

**12**  What fraction of 2 metres is 8 centimetres?

**13**  How many seconds in 5 hours?

**14**  Approximately, what is 5.9 squared?

Ask an adult to time you.

You need pencil and paper. Write only the answers.

**1**  £9.32 – £5.68.

**2**  Half of 672.

**3**  1200 ÷ 100.

**4**  64 + 93.

**5**  Write 120% as a decimal.

**6**  9 + □ + 17 = 50.

**7**  1.2 – 0.9.

**8**  Find the product of 1.8 and 5.

**9**  3 pizzas at £1.96 each cost … ?

**10**  If $a = 9$ and $b = 6$, what is $a - b$?

**11**  Add 13, 17, 21, 6 and 5.

**12**  Approximately, what is 28.1 ÷ 3.9?

**13**  30% of 60.

**14**  What is the cube root of 64?

**14**

# Signposts

Try these by yourself.

You need pencil and paper.

Each box represents a missing sign (+, –, × or ÷).

Can you find out what it is?

**a.**    (56 □ 38) □ 62 = 1116

**b.**    (2030 □ 35) □ 97 = 155

**c.**    650 □ (48 □ 35) = 50

**d.**    27 □ (13 □ 15) = 5265

Estimate and approximate
Use knowledge of number facts and times-tabl

# Sports shop

Do this by yourself.

You need pencil and paper.

Write a list of the sports items.

**a.** The price of each item will be cut by 15% in a sale.

Write its sale price next to each item.

**b.** Work out your change from £100 for each item in the sale.

**c.** Work out the total cost if you buy all the items in the sale.

Calculate change and total a shopping bill
Calculate percentages

# Owls

Start this on your own.

You need paper, a pencil and some scissors.

Make 10 number cards.

| 1 | 2 | 3 | 4 | 5 | 6 | 7 | 8 | 9 | 10 |

Arrange your cards in the shapes of the letters of OWLS.

Each line of the letter must add up to the same number.

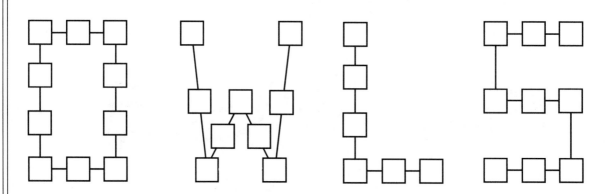

**1**   For O:   use 1 to 10.   Make each line add up to 18.

**2**   For W:   use 1 to 9.   Make each line add up to 16.

**3**   For L:   use 1 to 6.   Make each line add up to 13.

**4**   For S:   use 1 to 9.   Make each line add up to 15.

Sketch each solution but keep it secret.

Now ask your family to try.

Add several small numbers
Think logically to eliminate what won't wor

Ask an adult to read you these.
You need pencil and paper.  Write only the answers.

**1**  30% of 200.

**2**  Double 0.85.

**3**  157 minus 62.

**4**  0.72 times 1000.

**5**  Add 500 to 3600.

**6**  6¾ plus 3¾.

**7**  Add one tenth to 6.9.

**8**  Find the change from £50 for £32.60.

**9**  Write all the factors of 124.

**10**  What is the cost of 15 buns at 9p each?

**11**  Take 45 centimetres from 3 metres.

**12**  How many seconds in ¼ of an hour?

**13**  Find the difference between 5.1 and 4.9.

**14**  Write 45 out of 100 as a percentage.

You need pencil and paper.  Write only the answers.

**1**  It is 01:47 hours.  What time was it 2¼ hours ago?

**2**  Which is closest to 25 kilometres:  10 miles, 15 miles or 20 miles?

**3**  What are the next three numbers:  1,  4,  9,  16,  … ?

**4**  How many thousands in one million?

**5**  What percentage of £10 is £7.50?

**6**  6 times a number is 96.  What is the number?

**7**  Write $^{68}/_7$ as a mixed number.

**8**  $2r = 8$.  What is the value of $r$?

**9**  How many faces has a cylinder?

**10**  A quarter of a litre of milk is added to this jug.
How many litres in the jug now?

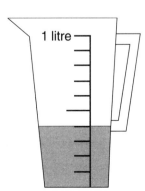

1 litre

# Treasure hunt

Do this by yourself.

You may need pencil and paper.

Josh is taking part in a treasure hunt.

He must start and end at point A.

There is a clue at each of the points A to I but he must stick to the paths shown.

What is the length of the shortest route he can take?

If there is also a clue somewhere on each path,

what is his shortest route now? How long is it?

Work out a strategy
Add a series of two-digit numbers

Ask an adult to time you.

You need pencil and paper. Write only the answers.

1   $34000 \div 1000$.

2   $1.6 \div 4$.

3   $2^2 \times 4^2$.

4   Write all the factors of 224.

5   $3004 - 1998$.

6   8.3 times 1000.

7   $(12 - 5) \times (14 + 6)$.

8   Write **103000** in words.

9   If $x = 8$, what is $x^2$?

10   Find the product of 0.1 and 3.

11   $19 + \square + 23 = 70$.

12   £7.56 + £8.29.

13   Simplify $2n + 7n$.

14   Approximately, what is half of 594?

# Use only these

Play with a partner.

Share a pencil and paper.

Draw this grid.

| 1 | 2 | 3 | 4 | 5 | 6 |
|---|---|---|---|---|---|
| 7 | 8 | 9 | 10 | 11 | 12 |
| 13 | 14 | 15 | 16 | 17 | 18 |
| 19 | 20 | 21 | 22 | 23 | 24 |

Use only these numbers and signs.

Take turns.

Say how to make a number on the grid.

The other player checks.

If you are right, mark the number with your initial.

If not, wait until your next turn.

You can mark each number only once.

Carry on until the grid is full.

The one who initials the most numbers wins the game.

### Change the rules

This time, use 3 and 7 instead of 4 and 9.

Make a 5 x 8 grid with the numbers 1 to 40.

Use knowledge of number bonds and times-tables
Think flexibly and work systematically

# Card tricks

Do this by yourself.
You need pencil and paper.

Imagine you have two cards.
Both cards have a number on each side.
You can place each card either way up.

a.   The **sum** of the two numbers showing could be 30, 31, 32 or 33.
     The numbers 14 and 17 are on the fronts of the cards.

     What numbers are on the backs?

b.   The **product** of the two numbers showing could be 84, 105, 108 or 135.
     The numbers 9 and 15 are on the fronts of the cards.

     What numbers is on the backs?

Make up two more puzzles like this.
Can your family solve them?

Practise addition and multiplication
Think logically to eliminate what won't work

# Home Maths Ages 11–12 Answers

**1**

| | |
|---|---|
| 1 | $\frac{1}{6}$ |
| 2 | 64 |
| 3 | £46.25 |
| 4 | 432 |
| 5 | 1016 |
| 6 | 1600 |
| 7 | 6 |
| 8 | 103 |
| 9 | 5.8 metres (or 5.8 m) |
| 10 | 10 |
| 11 | 47 |
| 12 | 1740 |
| 13 | $\frac{7}{8}$ |
| 14 | 14.35 |

**2**

1. $\frac{2}{3} + \frac{1}{2} + \frac{3}{4} = 1\frac{11}{12}$

2. $\frac{4}{5} + \frac{1}{2} + \frac{5}{6} = 2\frac{2}{15}$

3. $\frac{2}{3} + \frac{3}{4} + \frac{5}{6} = 2\frac{1}{4}$

4. $\frac{4}{5} + \frac{1}{2} + \frac{3}{4} = 2\frac{1}{20}$

5. $\frac{2}{3} + \frac{4}{5} + \frac{5}{6} = 2\frac{3}{10}$

**3**

| | | |
|---|---|---|
| 1 | SHE | 107 |
| 2 | PIE | 151 |
| 3 | HER | 162 |
| 4 | EWE | 206 |
| 5 | HIS | 63 |
| 6 | PEW | 173 |
| 7 | PIP | 118 |
| 8 | SEW | 140 |
| 9 | HISS | 73 |
| 10 | WHIP | 150 |
| 11 | CREW | 282 |
| 12 | WISH | 117 |

More words: HIS, HER, PIE, RIP, SIR, WICK, KISS, HIPS, PIER …

**5**

| | |
|---|---|
| 1 | $\frac{12}{100}$ or $\frac{3}{25}$ |
| 2 | 20 |
| 3 | 172 |
| 4 | 14 |
| 5 | 450 |
| 6 | –8 |
| 7 | 2500 |
| 8 | 5.8 metres (or 5.8 m) |
| 9 | 1 kilogram is more |
| 10 | 9500 |
| 11 | 170 |
| 12 | 24 |
| 13 | 2 |
| 14 | 150% |

**6**

| | |
|---|---|
| 1 | 27°C |
| 2 | 43 m.p.h. |
| 3 | $\frac{1}{4}$, $\frac{1}{2}$, $1\frac{1}{2}$, $1\frac{3}{4}$, 2 |
| 4 | 0.8 |
| 5 | 1.65 metres (1.65 m), or 165 centimetres (165 cm) |
| 6 | two (53 and 59) |
| 7 | 13 |
| 8 | 18 edges |
| 9 | 94.4 |
| 10 | £3.45 |

**7**

The 11 possible totals are:

$2\frac{1}{30}$  $2\frac{1}{20}$  $2\frac{1}{12}$  $2\frac{1}{6}$  $2\frac{4}{15}$  $2\frac{1}{2}$
$2\frac{9}{10}$  $3\frac{1}{3}$  $4\frac{1}{4}$  $5\frac{1}{5}$  $6\frac{1}{6}$

The 11 possible differences are:

$\frac{11}{30}$  $\frac{9}{20}$  $\frac{7}{12}$  $\frac{5}{6}$  $1\frac{1}{15}$  $1\frac{1}{2}$
$2\frac{1}{10}$  $2\frac{2}{3}$  $3\frac{3}{4}$  $4\frac{4}{5}$  $5\frac{5}{6}$

**8**

| | |
|---|---|
| 1 | 12.24 |
| 2 | 891 |
| 3 | 47 |
| 4 | $\frac{1}{10}$ |
| 5 | 5.2 |
| 6 | 5600 |
| 7 | 45 |
| 8 | 130 |
| 9 | £4.39 |
| 10 | $\frac{1}{6}$ |
| 11 | 698 |
| 12 | 1.8 |
| 13 | £15.60 |
| 14 | 26 |

**9**

4 unequal

6 equal

7 unequal

5 unequal

7 equal

8 equal

or

### 11

1. 7 pints
2. 12
3. 6
4. 13
5. £1.74
6. 32, 64, 128
7. 01:53 hours
8. 4 vertices
9. $7\frac{3}{8}$
10. 30 m.p.h.

### 12

1. 560
2. 6.3
3. 1003
4. 216
5. 492
6. 85
7. 18
8. 2100
9. 65
10. 112p or £1.12
11. 60%
12. $\frac{8}{200}$ or $\frac{1}{25}$
13. 18 000
14. 36

### 13

1. £3.64
2. 336
3. 12
4. 157
5. 1.2
6. 24
7. 0.3
8. 9
9. £5.88
10. 3
11. 62
12. 7
13. 18
14. 4

### 14

a. $(56 - 38) \times 62 = 1116$

b. $(2030) \div 35) + 97 = 155$

c. $650 \div (48 - 35) = 50$

d. $27 \times (13 \times 15) = 5265$

### 15

|  | price | change |
|---|---|---|
| Racket | £41.31 | £58.69 |
| Cricket bat | £57.97 | £42.03 |
| Sports bag | £21.25 | £78.75 |
| Football | £28.39 | £71.61 |
| Tennis balls | £7.65 | £92.35 |
| Ping pong bat | £10.20 | £89.80 |
| Roller skates | £21.08 | £78.92 |
| TOTAL | £187.85 | |

### 16

For example,

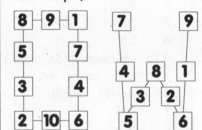
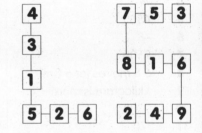

### 17

1. 60
2. 1.7
3. 95
4. 720
5. 4100
6. $10\frac{1}{2}$
7. 7
8. £17.40
9. 1,124   2,62   4,31
10. £1.35
11. 2.55 metres (2.55 m), or 255 centimetres (255 cm)
12. 900 seconds
13. 0.2
14. 45%

### 18

1. 23:32 hours
2. 15 miles
3. 25, 36, 49
4. 1000
5. 75%
6. 16
7. $9\frac{5}{7}$
8. 4
9. 3 faces
10. 0.65 litre (0.65 l), or 650 millilitres (650 ml)

### 19

Shortest route through each point is 207 metres: for example, AGBHCDIEFA

Shortest route over each path is 486 metres: for example, ABCDEFAGFIEIDHCHBGHIGA

### 20

1. 34
2. 0.4
3. 64
4. 1,224   2,112   4,56   8,28   14,16   7,32
5. 1006
6. 8300
7. 140
8. one million, thirty thousand
9. 64
10. 0.3
11. 28
12. £15.85
13. 9n
14. 300

## Card 21

Clues: for example,

| | | | |
|---|---|---|---|
| **1** | 9 – 4 – 4 | **13** | 49 – (4 x 9) |
| **2** | 4 x (9 – 4) – 9 – 9 | **14** | 4 x 9 – 9 – 9 – 4 |
| **3** | 4 x 4 – 9 – 4 | **15** | 4 x 9 – 9 – 4 – 4 – 4 |
| **4** | 4 | **16** | 4 x 4 |
| **5** | 9 – 4 | **17** | 44 – 9 – 9 – 9 |
| **6** | 4 x 9 – 9 – 9 – 4 – 4 – 4 | **18** | 4 x 9 – 9 – 9 |
| **7** | 4 x 4 – 9 | **19** | 4 x 9 – 9 – 4 – 4 |
| **8** | 4 x 4 – 4 – 4 | **20** | 4 x (9 – 4) |
| **9** | 9 | **21** | (9 – 4) x (9 – 4) – 4 |
| **10** | 4 x 9 – 9 – 9 – 4 – 4 | **22** | 44 – 9 – 9 – 4 |
| **11** | 4 x 4 – (9 – 4) | **23** | 4 x 9 – 9 – 4 |
| **12** | 4 x 4 – 4 | **24** | 4 x 9 – 4 – 4 – 4 |

## Card 22

**a.** Each card has 16 on the back.

**b.** 9 has 7 on the back, and 15 has 12 on the back.

## Card 23

**1** 38 centimetres
**2** 36, 45, 55
**3** five
  (101, 103, 107, 109, 113)
**4** 250 grams (or 250 g)
**5** 6 hours 4 minutes
**6** 155
**7** 8 faces
**8** 3/20
**9** 3
**10** £350

## Card 24

**1** 1188
**2** 1.6 or 1³⁄₅
**3** £19.14
**4** 210
**5** 634
**6** 10 100
**7** 70
**8** 1¼
**9** 1000 cm³
**10** 1 hour 44 minutes
**11** £7.84
**12** 3.3 centimetres (3.3 cm), or
  33 millimetres (33 mm)
**13** 8
**14** 11

## Card 25

**1** 9170
**2** 60
**3** 72
**4** 4.77
**5** 0.31
**6** 4
**7** £5.38
**8** 60
**9** £1.92
**10** 0.09
**11** 6a
**12** 12
**13** 3
**14** 3 000 060

## Card 26

| × | 25 | 42 | 58 | 37 | 61 |
|---|---|---|---|---|---|
| **40** | 1000 | 1680 | 2320 | 1480 | 2440 |
| **18** | 450 | 756 | 1044 | 666 | 1098 |
| **36** | 900 | 1512 | 2088 | 1332 | 2196 |
| **29** | 725 | 1218 | 1682 | 1073 | 1769 |

## Card 27

| | | |
|---|---|---|
| **a.** | 33 | BARGE |
| **b.** | 38 | CANOE |
| **c.** | 47 | SMACK |
| **d.** | 49 | KAYAK |
| **e.** | 51 | SKIFF |
| **f.** | 57 | YACHT |
| **g.** | 58 | LINER |
| **h.** | 77 | SLOOP |

## Card 28

| | | | |
|---|---|---|---|
| **a.** | $3^2 + 4^2$ | **k.** | $5^2 – 3^2$ |
| **b.** | $5^2 + 5^2$ | **l.** | $6^2 – 4^2$ |
| **c.** | $8^2 + 4^2$ | **m.** | $8^2 – 7^2$ |
| **d.** | $5^2 + 6^2$ | **n.** | $7^2 – 3^2$ |
| **e.** | $9^2 + 2^2$ | **o.** | $9^2 – 2^2$ |
| **f.** | $6^2 + 4^2$ | **p.** | $15^2 – 14^2$ |
| **g.** | $10^2 + 5^2$ | **q.** | $13^2 – 5^2$ |
| **h.** | $12^2 + 5^2$ | **r.** | $11^2 – 3^2$ |
| **i.** | $10^2 + 2^2$ | **s.** | $15^2 – 6^2$ |
| **j.** | $8^2 + 9^2$ | **t.** | $16^2 – 7^2$ |

## Card 29

**1** 134
**2** 3.36
**3** 0.64
**4** 8400
**5** 260
**6** 297
**7** 0.35
**8** 6.3
**9** 18
**10** 3
**11** £3.36
**12** 41p
**13** 9
**14** 45%

## Card 30

For example,

$$79 + 5\tfrac{2}{6} = 84\tfrac{1}{3}$$

(the 9 and 5, and the ²⁄₆ and ¹⁄₃, are interchangeable)

or

$$48 + 9\tfrac{2}{6} = 57\tfrac{1}{3}$$

(the 8 and 9, and the ²⁄₆ and ¹⁄₃, are interchangeable)

## 31

1. 2 kg
2. 54
3. 14
4. 5.8 km/min, or 348 km/h
5. 21, 34, 55
6. ⅕
7. 100
8. 7
9. 10 000
10. −9°C

## 32

or

## 33

1. 169
2. 1683
3. 23.75
4. 16
5. 3.9
6. 0.73
7. 6a
8. 0.32
9. 25
10. £12.51
11. 44
12. 1017
13. k = 3
14. ³⁄₅₀

## 35

1. 78
2. 95
3. 36
4. 747
5. 333
6. 532

## 36

1. 20 centimetres (20 cm)
2. 100 000
3. 0.12
4. 26 January
5. 29, 47, 76
6. 1100 miles
7. 1000
8. 3
9. 12 inches
10. £12.50

## 37

1. 2994
2. 0.9
3. 610
4. 65
5. 729
6. 0.36
7. 3½
8. 80
9. 7650
10. 210
11. 0.125
12. 3.05
13. 16 ounces (16 oz)
14. 292

## 38

1. 0.4
2. 9.8
3. 108
4. £13.50
5. 256
6. 0.7
7. £3.72
8. 29
9. 1.71
10. £15.80
11. 0.01
12. ³⁄₂ or 1½
13. 8.99
14. 465

## 40

a.
$$\begin{array}{r} 1\ 7\ 3\ 8 \\ \times\qquad 4 \\ \hline 6\ 9\ 5\ 2 \end{array}$$

b.
$$\begin{array}{r} 1\ 9\ 6\ 3 \\ \times\qquad 4 \\ \hline 7\ 8\ 5\ 2 \end{array}$$

c.
$$\begin{array}{r} 4\ 8\ 3 \\ \times\quad 1\ 2 \\ \hline 5\ 7\ 9\ 6 \end{array}$$

d.
$$\begin{array}{r} 1\ 5\ 7 \\ \times\quad 2\ 8 \\ \hline 4\ 3\ 9\ 6 \end{array}$$

## 42

1. 336
2. 0.54
3. 1 500 000
4. 3150
5. 0.06
6. 54
7. 343
8. 4
9. 1 metre
10. 175%
11. 125 000
12. 68
13. £3.48
14. 665

## 43

1. 18
2. 243, 729, 2187
3. 51.6, 5.66, 5.6, 5.06
4. £120.24
5. 2
6. 20.0
7. 7%
8. 100
9. 31
10. 1934

## 44

a. 132 × 45 = 5940
b. 93 × 86 = 7998
c. 37 × 49 = 1813
d. 336 × 43 = 14 448
e. 383 × 37 = 14 171

## 45

1. 180
2. £2.52
3. 200
4. 338
5. 225
6. 86
7. 43
8. 0.216
9. ⁸⁄₇ or 1⅐
10. 0.875
11. £13.84
12. 6.82
13. 12 060
14. a = 4

You need pencil and paper. Write only the answers.

**1** 950cm of tape is cut into 25 equal pieces. How long is each piece?

**2** What are the next three numbers: 10, 15, 21, 28 … ?

**3** How many prime numbers lie between 100 and 120?

**4** Which is closest to 8 ounces: 250g, 500g or 750g?

**5** How long is it from 22:41 hours to 04:45 hours?

**6** Add up all the whole numbers from 11 to 20.

**7** How many faces has a hexagonal prism?

**8** What fraction of £10 is £1.50?

**9** $5a = 15$. What is the value of $a$?

**10** A holiday to Disneyland costs £500. There is a 30% discount. What does the holiday cost?

Ask an adult to read you these.
You need pencil and paper. Write only the answers.

**1** 12 times 99.

**2** 8 divided by 5.

**3** Double £9.57.

**4** 50 plus 70 plus 90.

**5** Multiply 6.34 by 100.

**6** Add 400 to 9700.

**7** Divide 7 by 0.1.

**8** Write 125 per cent as a fraction.

**9** What is the volume of a 10 cm cube?

**10** How long is it from 22:16 to midnight?

**11** What is the cost of 8 pens at 98p each?

**12** Take 7 millimetres from 4 centimetres.

**13** 96 divided by 12.

**14** $4\frac{5}{8}$ plus $6\frac{3}{8}$.

Ask an adult to time you.

You need pencil and paper.  Write only the answers.

**1**  9.17 x 1000.

**2**  56 − 27 + 31.

**3**  $3^2 \times 2^3$.

**4**  5 − 0.23.

**5**  2.17 ÷ 7.

**6**  □ × 35 = 140.

**7**  £7.84 − £2.46.

**8**  If $n = 15$, what is $4n$?

**9**  6 tubes of sweets at 32p each cost …?

**10**  Multiply 9 by 0.01.

**11**  Simplify $4a + 2a$.

**12**  132 ÷ 11.

**13**  (56 + 34) ÷ (35 − 5).

**14**  Write in figures **three million and sixty**.

# Tables

Do this by yourself.

You need pencil and paper.

Copy and complete this multiplication table.

| x | 25 | 42 | | | 61 |
|---|---|---|---|---|---|
| 40 | 1000 | | | | |
| | | | 1044 | 666 | |
| | 900 | | 2088 | | |
| | | | | | 1769 |

Practise long multiplication and division
Work out a strategy

# All at sea

Do this by yourself.

You need pencil and paper.

The names of 8 kinds of boats are hidden in this puzzle.

To find them, follow a continuous track once through every square.

You can move up, down or sideways but not diagonally.

**Start here**

| K | C | S | F | F | I | H | C |
|---|---|---|---|---|---|---|---|
| L | A | M | P | S | K | T | A |
| I | N | E | O | O | L | S | Y |
| A | C | R | A | Y | B | A | E |
| N | O | E | K | A | K | R | G |

Each letter stands for a number. A is 1, B is 2, C is 3, and so on.

You will need to work out the rest.

Add up the numbers in the boats' names.

Which boats have these totals?

| | | | | | | | |
|---|---|---|---|---|---|---|---|
| **a.** | 33 | **c.** | 47 | **e.** | 51 | **g.** | 58 |
| **b.** | 38 | **d.** | 49 | **f.** | 57 | **h.** | 77 |

Add several one- and two-digit numbers

# Squares

Do this by yourself.

You need pencil and paper.

Some numbers can be written as the sum of two different square numbers.

For example,

$$34 = 3^2 + 5^2$$

Write each of these numbers as the **sum** of two squares.

| | | | | | | | | | |
|---|---|---|---|---|---|---|---|---|---|
| **a.** | 25. | **c.** | 80. | **e.** | 85. | **g.** | 125. | **i.** | 104. |
| **b.** | 50. | **d.** | 61. | **f.** | 52. | **h.** | 169. | **j.** | 145. |

Write each of these numbers as the **difference** of two squares.

| | | | | | | | | | |
|---|---|---|---|---|---|---|---|---|---|
| **k.** | 16. | **m.** | 15. | **o.** | 77. | **q.** | 144. | **s.** | 189. |
| **l.** | 20. | **n.** | 40. | **p.** | 29. | **r.** | 112. | **t.** | 207. |

Calculate, add and subtract squares
Work out a strategy

Ask an adult to read you these.

You need pencil and paper.  Write only the answers.

| | | | | |
|---|---|---|---|---|
| **1** | 56 plus 78. | **8** | Add one hundredth to 6.29. |
| **2** | 336 divided by 100. | **9** | Divide 9 by one half. |
| **3** | Multiply 64 by 0.01. | **10** | What is the cube root of 27? |
| **4** | Take 700 from 9100. | **11** | What is the cost of 7 ices at 48p each? |
| **5** | 4 times 65. | **12** | £3.28 shared equally among 8. |
| **6** | Subtract 8 from 305. | **13** | Write 9.09 to the nearest whole number. |
| **7** | Half of 0.7. | **14** | Write 4½ out of 10 as a percentage. |

# Digit sum

Do this by yourself.

You need pencil, paper and scissors.

Make nine number cards.

| 1 | 2 | 3 | 4 | 5 | 6 | 7 | 8 | 9 |

Arrange the cards to complete this sum. Use each card once.

$$\Box\Box + \Box\dfrac{\Box}{\Box} = \Box\Box\dfrac{\Box}{\Box}$$

Add mixed numbers
Think logically

You need pencil and paper. Write only the answers.

**1**  A 12kg log is split in the ratio 5:1. Is the smaller part 1kg, 2kg or 5kg?

**2**  How many days is it from June 13 to August 6?

**3**  A bottle of coke costs 35p. How many bottles can you get for £5?

**4**  A plane goes 58km in 10 minutes. What is its average speed?

**5**  What are the next three numbers: 1, 2, 3, 5, 8, 13, … ?

**6**  What fraction of 2 metres is 40 centimetres?

**7**  What is the square root of ten thousand?

**8**  $b^2 = 49$. What is the value of $b$?

**9**  How many hundreds in one million?

**10**  The temperature is −7°C. It falls by 8°C, then rises by 6°C. What is it now?

# Magic star

Do this by yourself.

You need pencil and paper.

Copy this diagram.

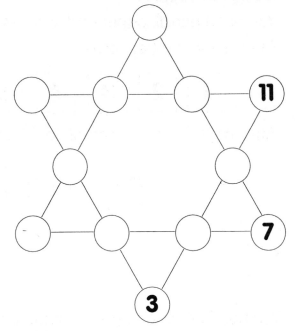

Use each of the numbers 1 to 12 once.

Write them in the circles.

Each line of four numbers must add up to 26.

Find two different ways to do it.

Add several small numbers
Think logically to eliminate what won't work

Ask an adult to time you.

You need pencil and paper.  Write only the answers.

| | | | |
|---|---|---|---|
| **1** | What is 13 squared? | **8** | $32 \times 0.01$. |
| **2** | $17 \times 99$. | **9** | If $x = 8$, what is $3x + 1$? |
| **3** | $23\,750 \div 1000$. | **10** | £4.83 + £7.68. |
| **4** | $4^3 \div 2^2$. | **11** | $13 + \square + 28 = 85$. |
| **5** | $4 - 0.1$. | **12** | $6007 - 4990$. |
| **6** | $4.38 \div 6$. | **13** | Solve $4k = 12$. |
| **7** | Simplify $3a + 5a - 2a$. | **14** | What fraction of 5 cm is 3 mm? |

# Lay them out

Three or four people can play.

You need a pack of playing cards.

Use cards 2 to 10 of each suit.

Take turns.

Shuffle the cards. Deal 12 cards face up in a line.

Secretly choose three neighbouring cards.

Without saying what they are call out their product.

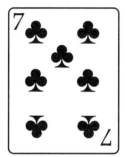  

**The product of 7, 3 and 8 is 168**
$$7 \times 3 \times 8 = 168$$

The other players try to find them …

or another line of three cards with the same product.

The first player to succeed keeps the three cards.

Return the rest of the cards to the pack and pass it to the next player.

Carry on until all the cards but three have been won.

For the last two rounds there will be fewer cards in the line.

The player who wins the most cards wins the game.

Play several games.

Practise short multiplication
Work out a strategy

# Who am I?

Do these puzzles by yourself.
You need pencil and paper.

## 1

Halve me.
Add 10 .
Find the square root.
The answer is 7.
Who am I?

## 2

I am a two-digit number.
I am three less than a multiple of 7.
I am one less than a multiple of 8.
I am not divisible by 13.
Who am I?

## 3

I have two digits.
I am a multiple of 4.
I am 27 less than my digits reversed.
Who am I?

## 4

My three digits add up to 18.
The first and the last are the same.
Their product is a square number.
Who am I?

## 5

I have three digits.
The first and the last are the same.
The product of my digits is a cube.
The sum of my digits is a square.
Who am I?

## 6

I have three digits.
I am the product of four primes.
The sum of these four primes is 30.
Each of my digits is a different prime.
Who am I?

Use knowledge of properties of numbers
Work systematically

You need pencil and paper.  Write only the answers.

**1**    Ceri cut 30cm of wire in the ratio 2:1.  How long in the longer piece?

**2**    One hundred times one thousand.

**3**    Which number is the largest:  0.08,  0.1,  0.12,  0.078?

**4**    What was the date 5 weeks before March 2?

**5**    What are the next three numbers:  1,  3,  4,  7,  11,  18,  … ?

**6**    Mum drives 13200 miles a year.  What is her monthly average?

**7**    What is the square root of one million?

**8**    $4m - 2 = 10$.  What is the value of $m$?

**9**    How many inches in 1 foot?

**10**    What is the cost of 1000 chews at 5p for 4?

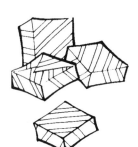

---

Ask an adult to read you these.
You need pencil and paper.  Write only the answers.

**1**    Subtract 8 from 3002.

**2**    90 divided by 100.

**3**    250 plus 360.

**4**    Divide 6.5 by 0.1.

**5**    Nine cubed.

**6**    0.4 times 0.9.

**7**    8¼ minus 4¾.

**8**    How many hundredths in 0.8?

**9**    Take 550 from 8200.

**10**    How many seconds in 3½ minutes?

**11**    Write one eighth as a decimal.

**12**    Half of 6.1.

**13**    How many ounces in 1 lb?

**14**    73 times 4.

Ask an adult to time you.

You need pencil and paper.  Write only the answers.

**1**  $0.5 \times 0.8$.

**2**  $0.98 \div 0.1$.

**3**  $3^3 \times 2^2$.

**4**  30% of £45.

**5**  What is 16 squared?

**6**  $700 \div 1000$.

**7**  £8.23 − £4.51.

**8**  If $b = 7$, what is $5b - 6$?

**9**  $8.55 \div 5$.

**10**  Four T-shirts at £3.95 each cost … ?

**11**  What is the square root of 0.1?

**12**  $^4/_5 \div {}^8/_{15}$.

**13**  $9 - 0.01$.

**14**  Multiply 15 by 31.

---

**39**

# Bull's eye

Play with a partner.

An adult with a calculator can act as referee.

Each player needs pencil and paper.

Take turns to go first.

The first player chooses a target number between 1 and 200.

Each player then chooses a **whole number**. They must divide 1000 by this number to get as close as possible to the target.

You can be under or over the target.

Now each of you do the division.

Round your answer to one decimal place.

The player to get closest to the target scores a point.

The first to get 10 points wins the game.

Estimate a quotient
Practise division to one decimal place

# Multiplying magic

Start this on your own.

You need paper, a pencil and some scissors.

Make 9 number cards.

Arrange the nine cards to complete these multiplications.

In each problem, use each card once and only once.

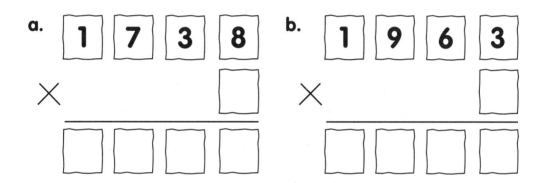

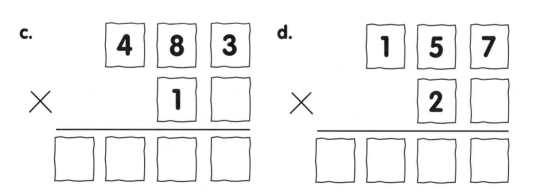

Record your solutions.

Now ask your family to try.

ractise multiplication
hink logically to eliminate what won't work

27

# Capture

Play with a partner.

You need paper between you.  You each need a different coloured pen.

Draw a grid of dots with fractions between them like this.

$$\begin{array}{cccccc}
\blacksquare & \blacksquare & \blacksquare & \blacksquare & \blacksquare & \blacksquare \\
+\tfrac{1}{2} & -\tfrac{5}{6} & +\tfrac{1}{3} & -\tfrac{7}{12} & +\tfrac{1}{4} \\
\blacksquare & \blacksquare & \blacksquare & \blacksquare & \blacksquare & \blacksquare \\
-\tfrac{3}{8} & +\tfrac{7}{12} & -\tfrac{1}{2} & +\tfrac{5}{8} & +\tfrac{2}{3} \\
\blacksquare & \blacksquare & \blacksquare & \blacksquare & \blacksquare & \blacksquare \\
+\tfrac{3}{4} & -\tfrac{5}{8} & +\tfrac{5}{12} & -\tfrac{3}{4} & -\tfrac{1}{2} \\
\blacksquare & \blacksquare & \blacksquare & \blacksquare & \blacksquare & \blacksquare \\
-\tfrac{1}{3} & +\tfrac{1}{2} & -\tfrac{7}{8} & +\tfrac{5}{6} & -\tfrac{5}{12} \\
\blacksquare & \blacksquare & \blacksquare & \blacksquare & \blacksquare & \blacksquare \\
+\tfrac{7}{8} & -\tfrac{1}{4} & +\tfrac{3}{8} & -\tfrac{2}{3} & +\tfrac{1}{6} \\
\blacksquare & \blacksquare & \blacksquare & \blacksquare & \blacksquare & \blacksquare \\
\end{array}$$

Take turns.

Join two neighbouring dots with a line.

Diagonal lines are not allowed.

If you complete a small square of dots, you capture the fraction inside.
Ring the fraction with your coloured pen and have another turn.

The winner is the player with the highest total in their boxes.

> **Change the rules**
>
> Put decimals in the boxes instead.

Add and subtract fractions
Think logically

Ask an adult to read you these.
You need pencil and paper.  Write only the answers.

1   How many hours in 2 weeks?
2   Multiply 0.6 by 0.9.
3   Write **1.5 million** in figures.
4   Add 350 to 2800.
5   60 divided by 1000.
6   91 minus 37.
7   Seven cubed.

8   Add one hundredth to 3.99.
9   Which is more: 1 yard or 1 metre?
10  Write 1¾ as a percentage.
11  What is one eighth of one million?
12  Divide 17 by one quarter.
13  What is the cost of twelve 29p stamps?
14  Multiply 35 by 19.

You need pencil and paper.  Write only the answers.

1   Dad divided 30 grapes in the ratio 3:2.  How many in the big portion?
2   What are the next three numbers:  1,  3,  9,  27,  81, … ?
3   Write in order, largest first:  5.06,  5.6,  51.6,  5.66.
4   What is the cost of 1002 bulbs at 12p each?
5   $3x + 1 = 7$.  What is the value of $x$?
6   Round 19.96 to one decimal place.
7   What percentage of £5 is 35p?
8   What is the cube root of one million?
9   Divide 899 by 29.
10  Gran was 18 in 1952.  When was she born?

# Missing digits

Try these by yourself.

You need pencil and paper.

Each box represents a missing digit. Can you find out what it is?

a.      $1\ \square\ 2 \times 4\ \square\ = 5940$

b.      $9\ 3 \times 8\ \square\ = 7\ \square\ \square\ 8$

c.      $3\ \square\ \times\ \square\ 9 = 1813$

d.      $\square\ \square\ 6 \times 4\ \square\ = 14\,448$

e.      $3\ \square\ \square\ \times\ \square\ 7 = 14\,171$

Practise multiplication and division
Think logically to eliminate what won't work

Ask an adult to time you.

You need pencil and paper. Write only the answers.

**1**    $33\frac{1}{3}$ per cent of 540.

**2**    £12.35 − £9.83.

**3**    $2^3 \times 5^2$.

**4**    6.76 ÷ 0.02.

**5**    What is 15 squared?

**6**    78 − 39 + 47.

**7**    Divide 1806 by 42.

**8**    Multiply 0.72 by 0.3.

**9**    $\frac{2}{3} \div \frac{7}{12}$.

**10**    Write $\frac{7}{8}$ as a decimal.

**11**    Add £3.62, £4.17 and £6.05.

**12**    7.1 − 0.28.

**13**    $402 \times 30$.

**14**    Solve $3a + 5 = 17$.

# Maze

Play with a partner.

You each need pencil and paper.

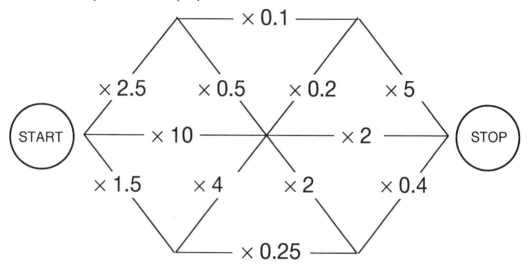

The first player chooses a whole number from 1 to 10 as a start number.

The second player chooses a whole number from 1 to 50 as a target.

Both think how to find a path from START to STOP.

Begin with the start number and keep multiplying by the number on the path.

Try to finish as close as possible to the target number.

You can travel along each path as often as you like.

Secretly jot down the path you will take.

When you are ready show each other your path.

The player with the smallest difference from the target number scores a point.

Play again.

Take turns to go first.

The winner is the first to get 5 points.

**Change the rules**

Choose from 50 to 60 to start.

Who can get closest to 100?

Practise multiplication of decimals
Think logically to eliminate what won't work

PUBLISHED BY THE PRESS SYNDICATE OF THE UNIVERSITY OF CAMBRIDGE
The Pitt Building, Trumpington Street, Cambridge CB2 1RP, United Kingdom

CAMBRIDGE UNIVERSITY PRESS
The Edinburgh Building, Cambridge CB2 2RU, United Kingdom
40 West 20th Street, New York, NY 10011-4211, USA
10 Stamford Road, Oakleigh, Melbourne 3166, Australia

First published 1998

Printed in the United Kingdom by Scotprint Ltd, Musselburgh

A catalogue record for this book is available from the British Library

ISBN 0 521 655501 paperback

Cover Illustration by Graham Round
Cartoons by Tim Sell

## This book covers:
- reading and writing large numbers
- recognising and using negative numbers
- rounding to the nearest 10, or 100
- rounding decimals to the nearest whole number or one decimal place
- adding, subtracting, multiplying or dividing whole numbers, fractions, decimals
- tables to 12 x 12
- multiplication and division by 10s, 100s, 1000s
- finding fractions, percentages and averages
- calculating ratios
- areas, volumes and perimeters
- reading scales
- knowledge of common 3D shapes
- solving simple problems